Why Ninjas Are Not In Fantasy Stories

By

Halden Mile

Cover art by: Michela
Art Cover design by: Rebeca Covers

PUBLISHED

BY

Table of Contents

Chapter 1
Some Ninja in Random Forest!

Once there was a Ninja who decided to explore a strange new land. This land wouldn't be new to the reader, nor would it be original. Why? It was populated by everything people would normally find in a fantasy story: Elves, Dwarves, Fairies you get the general idea. In short, anything in this land could have been plagiarized from every fantasy novel ever written.

(Now, dear reader, this is not an attempt to bad-mouth the entire Fantasy genre. After all, it is a remarkable genre. However, this tale should help explain why Ninjas are not present in it.)

Ahead, the Ninja could see a bridge that crossed into an enchanted forest of some kind. It was bound to be full of danger. Since the Ninja was. well, you know and anything dangerous seemed like fun, the Ninja went ahead and traveled to the nearby bridge.

At the bridge lived an evil, nasty troll with horrible drool, fangs, and all sorts of knots and warts on its hairy body. In short, there was nothing appealing about it. The troll had a habit of eating anyone who came near his bridge. Any human unlucky enough to approach had the unfortunate luck of being eaten.

On this day, however, the troll did not expect a visit from a Ninja. Thinking he would be just another easy target, the troll came out from under the bridge and threatened to eat the Ninja.

"Ye there!" the nasty troll bellowed, blocking the bridge. "I'm going to eat yeh!"

The troll spoke with a horribly fake Scottish accent. This went along with all the fake British accents spoken in every fantasy story.

Showing no fear whatsoever, the Ninja grabbed the troll by its horns.

"Hey, what are you doing!?" shouted the troll.

Its terrible accent vanished amid the shock. "Let me go!!"

The troll was being flung around by the Ninja, who had an impressively strong grip. In a brutal fashion, the Ninja succeeded in ripping out the troll's horns.

"My horns!!" exclaimed the troll, howling in pain as blood ran down its head. "My horns!! Your scumbag!! You ripped out my horns! Your scoundrel! You… filthy bag of manure!!"

(It should be noted that there are no swear words in fantasy novels unless written by George R. R. Martin and his books about a throne. Was it a toilet? We do not know exactly. Every fantasy novel usually has ye olde English.)

While the troll howled in pain, the Ninja created nun chucks from its horns. Then the Ninja used them to defeat the troll.

From out of nowhere, a minstrel appeared to sing:

"A brave warrior, in clothes droll,

Hath de-feated a terrible troll!

Impaling its deplorable body, Atop

A very tall tree!

For all the travelers to see!

Liberated from paying a crossing toll!"

Annoyed by this awful singing, it's no surprise the Ninja killed the minstrel. Afterward, he went on his merry way. The forest had big trees, which were common for a forest. (After all, a forest without trees would be like a desert without dunes.)

Little did the Ninja know, there were mischievous Pixies in the trees. The Pixies had a fetish for causing pranks to people who walked in the forest. When they saw the Ninja, they thought it would be fun to harass him.

Big mistake

The Pixies wasted no time messing with the Ninja. They yanked his blue gi and tabi boots, which made him very infuriated. He then pulled out his shuriken and threw every one of them at the Pixies. The torturous Pixies found themselves pinned to the thick trees in the forest where they oozed out bright orange blood.

When the Ninja was done, he continued onward into the forest. There, he saw a Talking Mushroom.

Since no one knew about the Talking Mushroom or why it was even in the forest, let alone why there was a Talking Mushroom to begin with the Ninja decided to see what the deal was regarding this "person." It should also be noted that the Ninja was very hungry at this point.

However, before he could find some food, the Talking Mushroom came up to the Ninja.

"Pardon me, sir," asked the Talking Mushroom. "I have a question to ask you!"

The Ninja rolled his eyes. "Oh, let me guess, you want to ask if I can obtain trinkets from some random mystical realm?"

"No, it's more along the lines of what do you call it when someone unwelcome intrudes on another one's path?"

The Ninja could not think as he was very hungry. He had forgotten to pack his sushi, rice balls, sake (that's SAH-KEE, not sake), sandwiches, spaghetti, fish, steak, fried chicken, cake, pizza, soup, fried squid, beer, salad, grapes, cherries, apples, soda, lingonberry spread, bread, toast, cheese, and the Ninja's very own lemonade. It was upsetting.

The Ninja's hunger soon overtook him. With a stupid Talking Mushroom asking a bunch of idiotic nonsense, the Ninja formulated a plan within his awesome mind. (Just watching this Talking Mushroom was enough to make anyone ponder how in this entire world such a thing came to fruition. Who was responsible for this?!)

"Well, in your case, it would have to be an annoying anthropomorphic plant who won't stop talking," answered the Ninja.

He then turned to the Talking Mushroom. "But I do have a question for you… whatever you are."

The Talking Mushroom was surprised. No one had ever asked him a question before. It was truly something that made his day; never once had he received a question from anyone.

He looked at the Ninja and said, "I never get questions from people, but I am curious. What do you wish to ask me?"

"Well…" the Ninja began. He couldn't finish as unbearable hunger clogged his thoughts.

After a moment, the Ninja finished his question. "What type of a person would you describe me as? I mean, you are a talking mushroom, but what am I?"

"I have no idea," answered the Talking Mushroom.

"You might be…"

"A HUNGRY NINJA!" the Ninja exclaimed.

He wasted no time pulling out his sword. This left the Talking Mushroom shaking to his very spores. In a matter of seconds, the Ninja sliced and diced the Talking Mushroom with his blade. The Talking Mushroom had no chance to react as it met its end at the hands of a Ninja who was very adept with a sword.

When the Ninja was finished (and the Talking Mushroom was dead), he built a fire, roasted the slices of the Talking Mushroom, and had himself a wondrous meal.

Once the Ninja had completed his lunch, he continued through the forest. He soon developed a thirst for water. Since he didn't have any beverages with him, the Ninja needed something to suffice for the time being.

Eventually, the Ninja found a brook. As he was very thirsty, he drank a good fill of the refreshing water.

There was no need to boil the water. Ninjas were immune to all sorts of things apart from rancid butter, that is.

As luck would have it, this brook was home to several Naiads and Undines. Both were angry and wanted to kill the Ninja out of revenge for the Pixies he had killed earlier.

However, little did the Naiads and Undines know that the Ninja seriously needed to use the bathroom.

As a result, the Ninja formulated a devious plot. When the Naiads and the Undines came close to him, the plucky hero dropped his pants and peed all over them. The Ninja emitted a stream of hot, yellow water flowing forth like a great river. The Naiads and Undines were helpless as they dissolved from being peed on.

Their last recorded words were: "Ack!! Yellow water!!"

Chapter 2

Big, Stupid Dragon, Mountain, Dwarves & Old Wizard Doing Something Bad!

After the Ninja had relieved himself, he washed his hands in the brook. Curiosity soon overtook him, and he set forth to follow the stream toward its source. Before long, the Ninja found himself at the base of a mountain with a tower perched on its peak. To reach the tower, he would have to pass through a cave at the base of the mountain.

The Ninja could have scaled the mountain, but the cave intrigued him. So, he decided to explore it.

However, inside the cave lurked a ferocious red dragon breathing fire, spreading its wings, swinging its powerful tail, and doing everything stereotypical of a villainous fantasy dragon. (So original. What's next some magical amulets?)

Everyone knows that nobody messes with a Ninja.

The Ninja drew his sword and reduced the dragon's scaly hide into raw materials, which he promptly sold to passing merchants who conveniently appeared at that exact moment. (How convenient! Does everything in this genre happen exactly when the characters need it most!?)

When the Ninja finished butchering the dragon, he decided to explore the cave further. Soon, he heard something troubling and decided to help. This proved that the Ninja did, in fact, have a good side.

He discovered a group of drunk Dwarves trapped in one of the mine shafts after the entrance caved in. (Considering all Dwarves ever do in fantasy is drink heavily and mine for minerals, it was bound to happen eventually.) Since the Dwarves were brave warriors it was a quality for the Ninja admired and he decided to save them.

Using his superhuman strength, the Ninja helped the Dwarves escape the collapsed mine shaft. But before they could make their way out, they were ambushed by a massive group of Orcs.

In one swift motion, the Ninja dispatched the entire Orc army, leaving the Dwarves in awe of his incredible speed. In fact, the Ninja was so fast that the Dwarves had only just begun to pull out their axes and flintlocks when the Ninja's killing spree was already over.

And that was just the first wave.

More waves of Orcs followed, and needless to say, the Ninja and the Dwarves had a lot of fun.

In gratitude, the Dwarves offered to help the Ninja find the exit leading to the tower on the mountain top. They also offered him some good mead. The Ninja, being a Ninja, did not get drunk. He was unstoppable against everything except rancid butter. The Dwarves, however, became so drunk that they drifted off to sleep.

"One day," vowed the Ninja, standing over his Dwarven allies, "you shall become great warriors with me."

Moving onward, once the Ninja found the tower, he entered and discovered more fiends in need of a beating.

One notable resident was an evil wizard who had a princess held hostage. In reality, the wizard was just a dirty old man wearing a bathrobe with stars and moons glued onto it. He occasionally cast weak spells with a little twig for a wand… which also served as a prostate examiner.

From the looks of things, the perverted old wizard was conjuring a spell. Even worse, it sounded more like his deranged fantasy of wanting to slide his gnarly hands into the princess's glittery pink underpants. Unfortunately for him, the princess wore a chastity garter.

The wizard tried everything to remove it: his gnarled hands, an ice pick, a sledgehammer, a jackhammer, and finally, a megaton nuclear warhead. (Don't ask where that came from, dear reader.) When none of it worked, the wizard remembered and he was a wizard. He could have just used magic. (Well, no duh.)

The wizard's ultimate plan was to sacrifice the princess or more accurately, assault her.

At that very moment, the Ninja burst into the tower. He saw what was happening and was infuriated.

The princess was hot, sexy, beautiful, and drop-dead gorgeous. In fact, there weren't enough adjectives in the dictionary to describe her. She had ample breasts, long flowing red hair, and perfect teeth.

(Considering the peasants stank to high heaven and their teeth were falling out, the princess was the finest woman in all the land.)

Even a Ninja could appreciate beauty, and seeing this old pervert threaten her was too much even by the Ninja's brutal standards.

"Who are you!?" the wizard exclaimed, spilling the contents of his Viagra. Blue pills scattered across the tower floor.

"Don't you see I'm in the middle of something here!? My flaccid, floppy penis is six feet under, so I need all the help I can get!"

"Hey, I didn't need to hear that! In fact, no one needs to know your cock is on life support!"

Our Ninja gestured toward a nearby rooster hooked up to a ventilator. Why it was there, no one knows.

"And as for who I am," the Ninja said, drawing his sword, "I'm a Ninja, and I'm here to kick your dirty old ass!"

"You mean him?" the perverted wizard pointed at a donkey that had wandered into the room.

The donkey did nothing but eat hay lying around for some reason. Where the hay came from, nobody knows. (Not even Halden Mile knows. He's the author of this story, so just go with it.)

The Ninja replied, "No, you."

In a split second, the Ninja leapt at the wizard and attacked with nun chucks, katanas, and every other Ninja weapon imaginable. It was, in short, the Ninja breaking the old man's hip. The wizard's magic was no match for him. To finish the job, the Ninja threw the wizard out of the tower and down the mountain.

The sound of the wizard splattering on the rocks below was music to the Ninja's ears.

The Ninja approached the princess but she suddenly vanished into thin air. She wasn't real. A blow-up doll would have been more convincing.

(Maybe she was an illusion from the wizard's magic. It's not like the pervert could explain anything now his face was acquainted with the rocks below.)

Suddenly, the tower began to crumble. But because the Ninja was… well, a Ninja, he hitched a ride on the donkey and escaped. Once safe, he used the rubble from the once-mighty tower to build a city.

The Ninja worked fast. Within hours, he transformed the ruins into a magnificent city with houses, stores, barns, a windmill, a couple of comic bookstores, a shopping mall, an airport (the Ninja invented flight as well), a film studio, a college, and a dojo to train new Ninjas.

When the city was complete, the Ninja set out to find people to live there.

His first destination: a nearby Elf city in the Plains of Vastness. It was here that the Ninja would find worthy souls to populate his new home.

Chapter 3
Brave Ninja with Elves, Elves, and More Elves!

Once the Ninja looked inside the Elf City, he was very impressed with what he saw. The architecture, crafted over centuries by Elves, was nothing short of magnificent. Tall, spiraling towers of elegant white marble pierced the sky. Grand statues of ancient heroes and heroines stood frozen in bronze, their gazes fixed on distant horizons.

Finally, there were the temples, majestic and serene, and surprisingly, a generous number of shopping malls, pizza joints, video game arcades, schools, warehouses, and pubs packed with Elves drunk on mead. These places were impressive, certainly, but in the Ninja's opinion, the ones he built were much better.

Whether or not the city had a dojo to train Elves in the way of the Ninja was unknown. What was known, however, was that the Ninja encountered Elves of every kind imaginable, from every walk of life and in all shapes and sizes. They spoke in every accent under the sun, especially those aggravatingly fake British accents.

There were poor Elves, rich Elves, fat Elves, thin Elves, young Elves, old Elves, jock Elves, punk Elves, country Elves, redneck Elves, vegan Elves, hippie Elves, beatnik Elves, goth Elves, straight Elves, gay Elves, Royalist Elves, Republican Elves, Democrat Elves, city Elves (which was redundant in this case), big Elves, little Elves, nerd Elves, dead Elves, medieval Elves, modern Elves, space Elves, nudist Elves (complete with black censor bars), White Elves, Black Elves, Latino Elves, Arabic Elves, Asian Elves, and even

Christmas Elves. (We would have added the Keebler Elves, but Kellogg's would have sent their lawyer Elves.)

Despite the city's stunning splendor, its breathtaking architecture and its wildly diverse inhabitants, it still paled in comparison to the Ninja's City of Awesomeness. Why? Because Ninjas are master builders, capable of creating anything from the simplest raw elements to the most complex devices capable of splitting atoms across the multiverse times infinity.

(In fact, legend has it that one Ninja once built a fully furnished, life-sized mansion using nothing but toothpicks and superglue.)

Let us also note that simply by the Ninja's presence alone, everyone flocked to him. Even the beautiful Elf girls who lived there, the only group of Elves worthy of the Ninja's attention and desire.

These Elves were not like the usual peasant stock, who were smelly, had teeth falling out, and were beyond gross. The Elf girls were much healthier and better looking than humans. Not even the fake princess from the tower was this good looking.

This was because Elf girls were of a higher quality in the way of slender, nimble bodies, lovely legs, and ample breasts. Lastly, the Elf girls had pointed ears protruding from their long, angelic golden hair. Some Elf girls were even redheads, brunettes, or had jet-black hair. (As if we did not already know what an Elf was. One must wonder if these same Elves run around either in green tights or half naked. Oh wait, we just said that.)

Regardless, even if these Elf girls were haughty and full of themselves, it did not matter. Every one of them was instantly enticed simply because a legendary Ninja had walked into their city. The Ninja did not even need to woo the minds of the beautiful Elf girls. They flocked to him simply because of his awesomeness. Some of the Elf girls screamed in excitement, and some even fainted.

Upon realizing he had won their hearts, the Ninja took the Elf girls back to his city on the mountain. Legend has it that the Elf girls were so enamored by the perfection of the Ninja's city that they accepted the Ninja as their shared suitor.

As a result, our Ninja hero spent his days telling the Elf girls tales of his exploits, his greatness, and making serious hot Elf love. (The loud, sensual moaning of the Elf girls was said to have been heard across the land. Their orgasms were rumored to cause slight changes in the weather.)

Yet deep down, the Ninja was bored out of his mind. He sought adventure because he did not like sitting around doing nothing. Sure, the Ninja was making serious Elf love, but it became very boring for him to keep rehashing the great stories of his greatness.

Eventually, the Ninja ran out of stories and resorted to displaying heroic feats of power. Whether it involved throwing shuriken, slicing fruit with his sword, or catching arrows with his hands, the Elf girls were in awe of his abilities.

Although this fascinated the Elf girls for a while, they eventually grew bored of these feats. And in turn, so did the Ninja. He needed something to bring back the thrill of living. He needed it fast, because life without adventure made him a very bored Ninja. He had been killing so many things left and right that he felt robbed of a good quest.

Then it came to him. Why not train those who traveled to visit him? After all, that would ease some of his boredom.

He sent the word out, and people from all walks of life came to him. Vagrants, the poor, adventurers, orphans, and even royalty flocked to train under the Ninja and learn his ways. Even the Elf women who lived with him would be trained.

Perhaps the best and most willing of the Ninja's students were the Dwarves who had fought with him against the waves of Orc armies. In a massive renovation, the Ninja converted his city into a City Fortress, complete with training grounds, a dojo, and traps for his enemies.

Everyone who came to learn the Ninja's ways had everything they needed to train and defend themselves. The main reason for this was to ensure that the Ninja's city, built from ruins, would be able to withstand any attack.

Regardless, the Ninja remained desperate for adventure. More than anything else in the world, he longed for it.

Chapter 4

Massive War: The Ultimate Ninja Fun Time!

One day, the Ninja received a message from a courier, which served as a blessed release from the pain of boredom. Word had spread of an evil Overlord who was preparing to conquer the land.

The villain had gathered a massive army of Goblins, Ghouls, Dragons, Orcs, Wizards (one of whom was the deranged pervert from before, now sporting a total hip replacement… just don't ask how he survived), Harpies, Skeleton Warriors, Giant Spiders, Snake People, and a mob of rabid, obese nerds. These nerds, complete with glasses, yellow teeth, and a bad case of acne, were furious that their latest role-playing session had been interrupted when their moms stormed into their basements to tell them it was way past bedtime. Their punishment was missing the latest comic convention. (Believe us, that would make ANY nerd upset. Even the hot, sexy cosplaying women who are impossible to date because they are either taken by that one lucky guy or into other women would feel the same way.)

There were even rumors that the evil Overlord was trying to recover several magical artifacts. These included gems, jewels, treasures, a toaster oven, a shaving razor, and a limited-edition ashtray which, for some reason, possessed great power. And of course, the free toy that came in the cereal box.

(What is this evil Overlord going to do with all this stuff? We can understand the gems, the jewels, and the treasures. But the toaster, the shaving razor, and the ashtray leave us utterly confused about what magical properties they could possibly have. Unless it has something to do with the evil Overlord's daily morning routine. What is even more unsettling being why anyone would create such magical artifacts for some deranged villain to acquire for world domination. Maybe we have missed something here, but unless somebody can drop us a line, we just don't know. Any help would be appreciated. And most of all, you can't go wrong with the free toy in the cereal box.)

Knowing that such a war would be exciting to our Ninja hero, he decided to take his newly trained fighters and embark on winning the war against the Overlord and his armies.

After traversing the land, people told the Ninja that the Overlord was ruling with an iron fist. However, his iron fist was really just an oven mitt. It turned out that the evil Overlord had been grilling his dollar store hot dogs when he decided to take over the land.

Why he wanted to rule the land, nobody knew for sure. Perhaps he just needed something to go along with his macaroni and cheese and cheap hot dogs.

The Ninja had a code of honor and would not tolerate the Overlord ruling over the land. Therefore, he decided to march to the Overlord's fortress and give the tyrant and his armies a massive fight. The Overlord sent his minions to stop him. There were Orcs, Harpies, Goblins, Skeleton Warriors, Giant Spiders, Snake People… you get the idea. In short, every single traditional minion of every single fantasy novel was there.

Legends spoke of these same minions wanting to protest the entire Fantasy genre. They wanted better representation and more positive portrayals. (There were even rumors of unions being organized in connection with this movement.) Sadly, that never came to pass because our Ninja hero dispatched them all in one clean slice of his sword. Thus, ending both their lives and their ability to complain about stereotypes.

(Well, what do you expect? There needs to be an army for the villains of fantasy stories and these guys usually fit the bill. Yes, it's stereotypical in their case, but hey, it's a living. So why complain?)

Finally, the Ninja came face to face with the Overlord inside his massive throne room. More accurately, it was the Overlord's kitchen. The kitchen might have looked intimidating had it not been for the window curtains made from doilies and lace.

On the windowsill sat a potted plant containing the rare flower of Some-Hard-To-Pronounce-Name-Without-A-Guide. As for its relevance to the story or plot, nobody has a clue.

The villain was sweating bullets in his thick armor. This was partly because he stood over the oven preparing orc soup. For those unfamiliar with the recipe, orc soup consisted of mud, dirt, old leather, dust balls, soiled underwear, urine, cow tongue, a princess's used tampons, and a big, nasty dragon turd. (Definitely something not appealing or appetizing.)

The second reason the Overlord was sweltering was his overwhelming fear of the Ninja. The Overlord was a coward who hid behind his powers and his charades, trying to make himself seem bigger than he was. The Ninja, however, feared nothing. He had traveled the world and could easily recognize when someone was pretending to be all high and mighty. This villain was one of those people.

In a desperate attempt to hide his fear, the Overlord spoke to the Ninja.

"WHO DARES ENTER MY THRONE ROOM?!" demanded the Overlord. His voice filled the room.

"You mean your kitchen?" the Ninja replied sarcastically. "Besides, this kitchen has the worst decorations ever."

The Ninja walked over to the refrigerator and helped himself to a Coke. As he chugged down the drink, the Overlord looked up from the nasty orc soup. Across the throne room, or kitchen rather, the Overlord saw the lone Ninja looking right back at him, a Coke in hand.

This shocked the Overlord as he knew all too well what would come next.

"I have come to challenge you!" exclaimed the Ninja. He discarded his empty Coke can and drew his sword.

That was the moment the Overlord finally realized whom he was dealing with.

"SO, YOU ARE THE ONE RESPONSIBLE FOR ALL THE CHAOS IN THE LAND?!" the Overlord thundered.

The Ninja might have caused a bit of a ruckus, but it was nothing compared to the chaos the Overlord had unleashed. That, and the Overlord's disgusting orc soup.

As he glared at the Ninja, the Overlord stepped away from the oven. His long, flowing black cape draped over his armored shoulders. In a theatrical display of power, he began shooting massive lightning bolts from his gauntlets. All of these were dodged by the Ninja with ease.

That was until the Overlord threw a plastic container from the countertop straight at the Ninja.

"Rancid butter!!" the Ninja shouted as the foul-smelling blocks of old butter hit him. It was torture, for the Ninja hated rancid butter with a passion.

In that moment, he felt an unexpected kinship with a Shaolin monk whose bald head had once been covered in spit tobacco.

Watching the Ninja react to the rancid butter made the Overlord erupt in laughter. Confident that his plan was now unstoppable, he turned his attention back to his kitchen. But things were about to turn around.

At that exact moment, a group of Elves and Dwarves rushed in. Upon seeing the Ninja covered in rancid butter, they immediately set to work cleaning him up. Using whatever they could find, from magic to soap rags, they removed every trace of the rancid butter from the Ninja's gi.

In retaliation, the Overlord decided to send out a giant. This giant was like any other giant, except for one thing: it had a terrible case of gas.

(Pheew! Someone crack a window!)

No one could withstand the terror of this giant's massive farting. The Elves and Dwarves were stunned by the sight and the smell of the creature.

"Yeesh! This giant is rather tainty!" they said, pinching their noses from the stench. "But we shall fight it!"

The Ninja was impressed with the resolve of his allies. For once, he stepped aside, curious to see what they would do.

When the giant was close, it let out a fart so powerful it cracked the walls, sending debris crashing to the floor. The rush of flatulence overpowered the Elves and Dwarves, causing several to pass out.

However, some of them managed to withstand the terrible smell. These would be the ones to bring down the giant once and for all.

Using their newly learned ninja abilities, the Elves and Dwarves constructed a massive catapult. Where they found the materials and tools to build it is a mystery. (Ninjas have an uncanny ability to produce anything at a moment's notice.) What mattered was that when they finished, they had built such a magnificent catapult that it brought a tear to the Ninja's eye.

"I am so proud of you all," the Ninja said, unable to contain his admiration. "May it serve us well in this battle."

As he spoke, the giant charged forward, hoping to knock out the remaining Elves, Dwarves, and the lone Ninja.

When the giant was close enough to unleash another devastating fart, the Elves and Dwarves launched the catapult's payload of pink bismuth straight into the giant's mouth.

The giant collapsed to the ground before them. It convulsed for a moment as the bismuth coated its throat and stomach. (Bismuth may have an acquired taste, but it works.) Its lips and throat turned pink as the medicine flowed down its esophagus.

After a few tense moments, the giant stood up again, shaking its head. The fiery sensation in its stomach was gone. No longer suffering from its terrible gas, the giant decided it had no reason to serve the Overlord anymore.

Before leaving, it walked over to the Elves, Dwarves, and the Ninja, bowing its enormous head.

"Thank you," said the giant. "I needed that."

Angered by the loss of his super weapon, the Overlord jumped down and challenged the Ninja to a sword fight. The Ninja loved this idea. It would be the perfect test of his skills. Finally, he had found something truly dangerous and fun.

The Ninja and the Overlord clashed in a frenzy. Their swords rang out in the heat of battle as hero and villain fought with all their strength. No matter what the Overlord tried, he was no match for the Ninja. Even with his magic and summoning abilities, he could not win.

Seizing the perfect moment to strike, the Ninja unleashed a devastating flurry of swords, nunchakus, shuriken, bombs, punches, and kicks. No one could have hoped to avoid such an attack.

When the Ninja was done, all that remained of the Overlord were shards of rusty armor. (Just one look at that armor would warrant a tetanus shot.)

Then, for the second time, the Ninja had to escape from a collapsing building.

To show once again how resourceful he was, the Ninja used the rubble from the fortress to build yet another city.

Unlike most fantasy novels, we will not bore you with long, wordy details. All you need to know is that it was an awesome city filled with even more Elf girls living within its walls. (When we say moving, we mean relocating inside the city, not trapped between slabs of stone. We would never advocate such a thing. Everyone loves Elf girls. Especially our Ninja hero.)

With peace restored to the land, the Ninja spent the rest of his days between the two cities he built from the ruins of tyranny. To keep boredom away, he offered his services to anyone in need, training anyone who wished to learn the ways of the Ninja.

Now you know why Ninjas will never appear in any fantasy stories. They are just too legendary. Yet one question remains. What in the hell happened to the rooster on the ventilator?

ABOUT THE AUTHOR

Halden Mile, based in North Florida, is a versatile artist whose talents span amateur acting, writing, YouTube blogging, ASMR, security work, and cosplay. With a captivating on-screen presence, sharp storytelling skills, and engaging online content, Halden embodies creative range and dedication. Whether bringing characters to life, ensuring public safety, or exploring the world of cosplay, he blends professionalism with imaginative flair.

A passionate fan of anime and film, Halden draws inspiration from the worlds he loves. Known for his wit and sharp sense of satire, he has a gift for weaving humor into unexpected corners of his stories. Born with an insatiable curiosity and a deep love for words, he embarked on a literary journey that has allowed him to explore multiple genres and captivate readers with his distinctive narrative voice.

Beyond his writing, Halden continues to be a dynamic creative force, leaving an indelible mark on the entertainment landscape.